Luiz Gustavo Batista Ferreira

Medicinal Species: Concepts and Microbiological Quality

Luiz Gustavo Batista Ferreira

Medicinal Species: Concepts and Microbiological Quality

Agronomy Course Conclusion Work

ScienciaScripts

Imprint

Cover image: www.ingimage.com

This book is a translation from the original published under ISBN 978-620-6-76198-3.

Publisher:
Sciencia Scripts
is a trademark of
Dodo Books Indian Ocean Ltd. and OmniScriptum S.R.L publishing group

120 High Road, East Finchley, London, N2 9ED, United Kingdom
Str. Armeneasca 28/1, office 1, Chisinau MD-2012, Republic of Moldova, Europe
Printed at: see last page
ISBN: 978-620-8-17451-4

MEDICINAL SPECIES: CONCEPTS AND MICROBIOLOGICAL QUALITY

COURSEWORK IN AGRONOMY

SUMMARY

The use of medicinal plants has been widespread throughout the world since the dawn of humanity, when the ancestors of the human species used plants as food and medicine. Their use is of great importance, especially in low-income populations without access to health resources. The aim of this study was to ascertain the phytosanitary quality of medicinal species sold in the city of Londrina. The medicinal plants were chosen on the basis of the structures that secrete or store the active ingredients, such as leaves, bark, flowers and fruit. The plant species chosen for the phytosanitary quality analyses were: chamomile (Chamomilla recutita L. Rauschert), mint (Mentha piperita), fennel (Pimpinella anisum L), hibiscus (Hibiscus sabdariffa L.), cinnamon (Cinnamomum zeylanicum Blume) and melissa (Melissa officinalis L.). The plant materials were purchased from five commercial outlets in the city: a street market (site 1); natural products shops (sites 2, 3 and 4) and a homeopathic pharmacy (site 5). The species purchased were transported to the laboratory. They were then placed in sterilised Petri dishes with BDA (potato dextrose agar) culture medium and 40g of each plant species divided into four different spots on the dish, and kept at a constant temperature of 25°C for seven days in the absence of light. Two replicates were carried out for each species and marketing point. Microorganisms that are harmful to human health were identified, such as Aspergillus flavus, Aspergillus niger, Aspergillus ochraceus, Fusarium sp, Nigrospora sp, Penicillium sp and Rhizopus sp. The six medicinal species bought in the Londrina shops were of poor phytosanitary quality. There was a high incidence of fungi harmful to human health, such as Aspergillus flavus and Penicillium sp, in the six medicinal species collected from the five commercial outlets.

Key words: *Medicinal plants. Fungi. Bacteria. Contamination.*

SUMMARY

1.INTRODUCTION 4

2.DEVELOPMENT 7

3.MATERIAL AND METHODS 20

4.RESULTS AND DISCUSSION 22

5.CONCLUSIONS 30

REFERENCES 31

APPENDICES 42

1. INTRODUCTION

Medicinal, aromatic and spicy plants have always been present throughout the evolution of the human species. The history of the use of medicinal plants shows us that, throughout this evolution, plants were the first therapeutic resources used because of man's own needs. Resorting to the healing virtues of certain plants is one of the first manifestations of this, marking an ancient and deep desire to understand and use nature as a therapeutic resource, and the first writings on plants with medicinal principles report their importance in magic and medicine ceremonies. They were used by various peoples, such as the Chaldeans, Babylonians and especially in Ancient Egypt, where recipes were made for the use of medicinal plants to cure the population's ailments. However, during the time of these ancient cultures, the medicinal properties of medicinal plants were attributed to mysticism, gods and the supernatural power of certain plants that could cure the illnesses and discomforts of the population. As studies on medicinal plants have progressed, it has been realised that the real reason why certain plants are able to cure ailments is called the active principle, whether it consists of a single substance in the plant or a set of substances that act synergistically to form a phytotherapeutic complex. These substances can be used either within the plant itself in the form of homemade preparations, such as teas, tinctures and powders, or in the form of a pure compound isolated from the plant, such as by the pharmaceutical industry. The medicinal plant, when well chosen and used correctly, only differs from the industrial medicine made from the isolated substance by its packaging and the colouring, flavouring and preservative substances that accompany the active ingredient in the final medicine (LORENZI, 2008). Thus, there has been an advance in the perspective and methods of utilising plants for medicinal use. According to the World Health Organisation (WHO), 80% of the world's population uses medicines derived from medicinal plants, and in Brazil research shows that more than 90% of the population has used a medicinal plant (ABIFISA, 2012). Brazil's rich plant diversity has contributed to the utilisation of medicinal plants being considered a strategic area for the country. According to Batalha et al. (2007), the country contains around 23% of the plant species catalogued by science. However, according to Veiga Júnior et al. (2005), there are serious problems in this context, for a number of reasons. Among them are the misuse of medicinal plants to perform cures (self-medication), wrong dosages which can make the active ingredients toxic to humans, inadequate handling in the field and lack of supervision by specialised bodies at points of sale, especially in emerging countries like Brazil, where the population that uses these products hardly values scientific knowledge and new discoveries about the use of medicinal plants, although there are also cases recorded in developed countries like Germany and the United States of America. In addition to the aforementioned problems related to the use of medicinal plants, there is another fundamental one, namely the phytosanitary quality of plants used in popular medicine, which are commercialised with pathogens that have a high potential to cause health problems for consumers. There is a suspicion that several centres of commerce in cities in Paraná, such as Londrina, sell plants with fungi that are harmful to human health, such as Aspergillus and Penicillium. These pathogens are considered storage fungi and their presence indicates poor quality or problems in the conservation of plant organisms. They can cause physico-chemical alterations in plant tissues, in flavour causing loss of lipids, alterations in the metabolism of the vegetable, loss of carbohydrates and proteins as well as an increase in fatty acids, compromising essential therapeutic properties, which result in harm to the human organism. In addition, Aspergillus and Penicillium, along with Fusarium, have species producers of mycotoxins (toxic substances potentially harmful to human health). In commercial areas, mainly street markets and shops, contaminated medicinal plants are sold, which can trigger a series of allergic reactions, jeopardising the health of the user. This is made worse by the lack of supervision of products, from the way they are grown to the conditions in which they are stored and sold to the end consumer. Although there are specific parameters for the production and trade of herbal medicines, these factors have occurred and have worried health professionals and the scientific community, since they jeopardise the use of these plants in the form of tea (MELO et al., 2007). The aim of this study was to assess the phytosanitary quality of: chamomile (Chamomilla recutita L. Rauschert), melissa (Melissa officinalis L.), fennel (Pimpinella anisum L.), cinnamon (Cinnamomum zeylanicum Blume), mint (Mentha piperita) and hibiscus (Hibiscus sabdariffa) sold in the municipality of Londrina - PR.

2. DEVELOPMENT

The Importance, Use And Study Of Medicinal Plants In The World Brazil.

The medicinal plant constitutes a therapeutic unit in which all the active principles form a phytocomplex that interacts with each other and with other apparently inactive molecules, which represents the integral pharmacological unit of the medicinal plant (JORGE, 2009). For Duarte et al, 2004, the biological activity of medicinal plants has been the subject of intense scientific research. Higher plants and aromatic plants are widely used in folk medicine, as they have a broad spectrum of activity and proven inhibition against bacteria and fungi, which means that medicinal plants are used in the formulations of medicines marketed by the pharmaceutical industry (MATOS and LORENZI, 2002). Most of these properties are conferred by products of secondary metabolism, such as terpenoids and phenolic compounds, and are present on a large scale in nature (SAAD et al, 2005).

Knowledge of medicinal plants often symbolises the only therapeutic resource for many communities and ethnic groups. They have been traded since ancient times and arrived in Brazil with the Portuguese colonisation (FIGUEIREDO, 2006). Even today, in the poorest regions of the country and even in large Brazilian cities, medicinal, aromatic and spicy plants are sold in open-air fairs, popular markets and found in residential backyards (MACIEL et al., 2002).

Martins (2003) highlights important data from the WHO: 80 per cent of the world's population has used tea, while in Brazil it is 90 per cent. The use of these plants has been stimulated, in part, by the industry's growing demand for new natural sources of medicines and, on the other hand, due to the side effects caused by synthetic drugs, which has stimulated the use of medicines of plant origin or, in many cases, because represent the only source of medicines, especially in the most isolated and remote places (BERG, 1993), and as a response to immediate health problems (DEFILLIPS, 2001). For Brito et al. (2009), the use of medicinal plants in curing diseases is demonstrated by the various examples cited in the literature regarding the use and efficacy of drugs of plant origin, such as studies using plants such as thyme, sage, oregano and coriander (PRUTHI, 1980), cloves (HITOKOTO et al., 1980; MABROUK; El-SHAYEB, 1980), cinnamon, rosemary, lavender (DAVIDSON, 1997), cumin, pepper, caraway (BEUCHAT; GOLDEN, 1989), garlic, onion (ABDOU et al., 1972; GANDHI; GHODEKAR,1988), basil, turmeric, marjoram and aniseed (HITOKOTO et al., 1980) in inhibiting human diseases. Ferreira (1998) categorises medicinal plants in general terms as plants that have biological activity with one or more active principles useful for health. They are plants that have a pharmacological action that is beneficial to the human organism (MARTINS et al., 1995). They are of direct importance in phytotherapy and are also raw materials for the pharmaceutical industry (KARL et al., 1997). The use of medicines, supplements and teas based on these plants is characterised as phytotherapy which, according to Sanguinetti (1989), has numerous advantages when used rationally and sensibly to treat ailments of the human body. Classical phytotherapy consists of the search for a cure for illnesses through the use of medicinal plants. Often, it is not an option for the population but the only resource (CARRICONDE, 2002). This practice has been widely disseminated from generation to generation, in the form of teas, infusions and licks (BRITO et al, 2009). The use of medicinal plants in Brazil has been used by native healers since before the beginning of Portuguese colonisation (MING et al., 2009). With regard to the empirical knowledge of popular phytotherapy, Morgan (1989) in his work stated that this practice augments the scientific method through trial and error. The WHO refers to medicinal plants as "any plant that contains, in one or more organs, substances that can be used for therapeutic purposes or that are precursors of semi-synthetic drugs". The difference between a medicinal plant and a herbal medicine lies in the preparation of the plant for a specific formulation, which characterises a herbal medicine (VEIGA JUNIOR et al, 2005). According to Duarte (2006), in the past, phytotherapy was more widely adopted by the poor in rural or urban areas, due to its easy availability and lower costs. Today, the use of plants as a source of medicines is prevalent in developing countries such as Brazil, as an alternative solution to health problems. The use of medicinal plants is well established in some continents such as Asia, Latin America and Africa. According to Martins et al. (2003), the use of plants in Brazil to treat illnesses is fundamentally influenced by indigenous, African and European culture. The Indians used phytotherapy from a mystical point of view. The European influence began with the arrival of the first priests of the Society of Jesus in Brazil. Traditional knowledge about the ecology and management of medicinal plants is fundamental for the rational and non-predatory utilisation of natural resources (CASTELLANI, 2003). Brazil has the greatest biological diversity in the world, with a rich flora, arousing the interest of international scientific communities in the study, conservation and rational utilisation of these resources. According to Plotkin (1991), Brazil is the country with the largest number of spermatophyte species in the world. Brazilian biomes contain some of the greatest diversity in the world (MENDONÇA et al, 1998), many of which have food and especially medicinal value (ALMEIDA et al, 1998). According to Duarte (2006), several Brazilian biomes have been used as natural drugs. In the medical sector, tropical plants provide material for the production of analgesics, tranquillisers, diuretics, laxatives and antibiotics, among others. The worldwide commercialisation of medicinal products of plant origin is on the increase (MEYERS,1983; PRINCIPE,1985), since there are treatments and synthetic medicines that are too expensive for the user. The science of ethnobotany is considered to be a new branch of science (HAMILTON et al., 2003) and Ford (1978) considered it to be the scientific study of the direct interaction between the human species and plants in dynamic systems (ALCORN, 1995) and has a broad interdisciplinary scientific character (ALBUQUERQUE, 2009). For Schultes and Reis (1995), ethnobotanical knowledge has existed since the origin of the Homo sapiens species. The term ethnobotany was first used in 1895 by Harshberger, who although he didn't define it, pointed out ways in which it could be useful for scientific research. Since then, various definitions can be found for ethnobotany. Among the most recent are: "discipline that deals with the study and conceptualisations developed by any society regarding the plant world" - it encompasses the way a social group classifies plants and the use it makes of them; "true scientific botany geared to the habitat and use of a specific ethnic group" - being carried out by someone trained in scientific botany, who would make correspondences between Western and local scientific classification; "botanical science that has a specific ethnicity" - sees the culture of a society as everything that someone has to know or believe in order to operate in a way that is acceptable to its members (MACIEL et al., 2002). The ethnobotanical study consists of evaluating the interaction This is done through surveys in traditional communities on the use of plants in home pharmacopoeia and in the domestic economy. One of the pioneers in ethnobotanical studies, Richard Evans Schultes, dedicated to researching the flora of tropical America, reports numerous Brazilian plant species used by caboclos and indigenous people in the Amazon (MARTIN, 1995). Various ethnobotanical studies have been carried out in Brazil and around the world, seeking to understand the folk medicine of traditional and/or contemporary peoples and the ways in which this knowledge is organised, They are also looking for plants that actually have a therapeutic activity and that consequently make it possible to discover new drugs (COUTINHO et al., 2002). These studies also make it possible to infer the efficacy of products that reach the market for the production of teas, syrups, creams and others (ALEXIADES and SHELDON, 1996). The study of the uses of medicinal plants must take into account the social and cultural context in which these uses are embedded

(HERRICK, 1983; ELISABETSKY, 1986; ETKIN, 1990; MACIEL et al., 2002).Many plants frequently used by local populations have not yet been studied or their active principles identified in order to validate them as medicines or to utilise them economically (BERG, 1993). Even so, many plants are used and commercialised today and Brazil, one of the most biodiverse countries in the world, is proving to be an important potential supplier of such a valuable resource as medicinal plants. Examples of valuable plants from Brazil include the indigenous Curare or Dedaleira (Digitalis purpurea) used to prepare tea against dropsy, caused by heart failure before the action of Digitalin on the heart muscle was discovered, Casca dánta (Drimys brasiliensis) with stomachic properties, Quina (Cinchona calisaya) used to cure malaria, Ipecacuanha (Cephaelis ipecacuanha) used to treat diarrhoea, amoebic dysentery, chronic catarrh, haemorrhage and asthma, and Sapucainha, (Carpotroche brasiliensis) with scientifically proven anti-inflammatory effects and whose oil extracted from the seed is used to treat leprosy (COUTINHO et al., 2002).The consumption of herbal medicines has increased considerably over the last two decades, both in developed and developing countries. In Europe alone, the market for herbal medicines is worth around 7 billion dollars a year, with Germany accounting for 50 per cent of this figure (ABIFISA, 2012). But in Brazil, Despite the importance of this market, there is no official data on the turnover of the herbal medicine industry. It is estimated to be worth around 1 billion reais/year (ABIFISA, 2012) with a tendency to grow more and more. Due to problems related to quality and regularity of supply, most pharmaceutical companies, 70% of which are transnational corporations based in Brazil, prefer to import their raw materials. Therefore, although there is a demand for medicinal plants and the potential to meet it, there is a lack of preparation on the part of farmers/gatherers to meet the demands of the market and the production chain in general.

Commercialisation of Medicinal Plants

According to Nunes et al. (2003), the trade and use of medicinal plants is well known and discussed in Brazil and around the world. The market caters for consumers of this type of product in different ways, including commercialisation by companies, markets and herbalists (BRANDÃO; FREIRE; VIANNA-SOARES, 1998a). In different countries, folk medicine using plants is widely practised by herbalists and small herbalists, and is expanding rapidly. There are also technically elaborated products made from medicinal plants, such as herbal medicines, a market that has also been expanding (CALIXTO, 2000; SUZUKI, 2002). The WHO has described the profile of this user, revealing that the majority of the world's population, living mainly in developing countries, does not have access to modern medicine and, consequently, to synthetic and herbal medicines, and therefore resorts to folk medicine (FARNSWORTH et al., 1985), which is available, among other forms, as products marketed by herbalists or extracted from sources. close to their place of residence. Various socio-economic factors have been pointed out to explain this phenomenon, ranging from the high cost of medicines to fads (BRANDÃO; FREIRE; VIANNA-SOARES, 1998b). The products marketed should respect the criteria of suitability for health and meet the safety and efficacy binomial, as is required of other products. What we have seen, however, is poor quality medicinal plants on the market, a lack of pharmacovigilance (BRANDÃO;FREIRE;VIANNA-SOARES,1998b) and the use of species that have not been fully evaluated from a pharmacological and toxicological point of view. In Brazil, this occurs mainly with native plants (RITTER et al., 2002), which are the most commercialised because they are obtained extractively. According to Barbosa et al. (2010), the lack of good practices in the production and commercialisation of medicinal plants and herbal medicines, combined with the absence of pharmacovigilance, contributes to significant drops in the quality of herbal medicines. In order to maintain the quality of a herbal medicine, it is necessary to follow appropriate practices in identifying, cultivating, collecting, handling, drying, storing and transporting them, as indicated in ANVISA Resolution RDC 306 of 07 December 2004, which is not the case with products sold by "raizeiros".

Pathogens harmful to humans present in commercialised plants.

Medicinal, aromatic and spice plants, as well as other plants of economic interest, can be affected by phytopathogens such as rusts, leaf spots, mycelium and oidium (RUSSOMANO and KRUPPA, 2010; LIMA et al., 2004). These diseases reduce the quality of medicinal plants and crop yields in general (CUMMINS and HIRATSUKA, 2003; HAWKSWORTH et al., 1995; AGRIOS, 1997; URBEN et al., 1987;SOUZA et al., 1995), with the main climatic factors contributing to for the incidence of plant diseases are temperature and high relative humidity in general (MELCHING et al., 1989; ALVES et al., 2006; TSCHANZ, 1984; DEL PONTE et al., 2006). Russomano and Krupa (2010) state that fungal diseases of the aerial parts of medicinal, spicy and aromatic plants should preferably be controlled by using resistant varieties and by using inoculum-free seeds, seedlings and propagating material in disease-free locations. The use of healthy, high vigour seeds prevents the introduction of pathogens into the beds, nurseries and the field and, when sown at the right depth, they enable the plant to emerge and grow quickly and remain susceptible to causal agents for less time. When there are no resistant varieties, the use of cultural control techniques can help to reduce the damage caused by diseases. In this context, the use of pesticides (fungicides) should be avoided due to the impact they have on the environment and human health through the pollution of their chemical residues. In addition, there is not enough information on the effect of fungicides on the metabolism of active, therapeutic and spicy ingredients, nor on the presence of toxic residues in extracts used for medicinal and culinary purposes. Nowadays, various innovative strategies have been employed to control diseases in agriculture, such as biological control using Trichoderma, the use of plant extracts, the elimination of cultural remains, crop rotation (with crops that are not host to the pathogens), balanced fertilisation, control of invasive species, appropriate and recommended spacing, growing plants in places with well-draining soil and pruning tree and shrub branches (RUSSOMANO and KRUPPA, 2010).

However, there are as yet no scientific studies or records of which prove that these diseases affect the humanorganism.Medicinal, aromatic and spicy plants can be hosts for agents harmful pathogens such as fungi of the genus Aspergillus, of the species fumigatus or niger (filamentous fungus), Penicillium and Fusarium, and these microorganisms can be marketed together with medicinal plants and the final consumer, through teas, for example, can ingest these fungi causing health complications through these microorganisms (TORTORA et al, 2000). The increase in the consumption of tea is worrying considering that the trade in these products is poorly supervised, thus posing various risks to the consumer, the main ones being misidentification of the plant species, incorrect indications for use, the use of poor quality plants and toxic species (ESPÍNOLA, 1997). According to Oliveira et al. (1997), Correa Junior et al. (1994) and Matos (2000), contamination of medicinal species by fungi can lead to alteration and/or destruction of the active ingredients, thus causing loss of safety and efficacy in use, as well as representing a risk due to the production of toxic substances (mycotoxins), thus making them unfit for consumption, regardless of the level of contamination. Aflatoxins, secondary metabolites produced by various species of fungi of the genus Aspergillus, are mycotoxins of great medical and health interest due to their high occurrence and high toxicity, causing acute or chronic poisoning, with hepatoxic, mutagenic, teratogenic and carcinogenic properties (ROITMAN et al, 1988) and are thermostable with a melting point of around 269°C (CARVALHO et al, 2010).

According to the International Agency for Research on Cancer (IARC), they are classified as a human carcinogen. According to Rosa (2012), there is morphological evidence in fungi that characterises the production of aflatoxins, such as the presence of blastoconidia, chlamydoconidia and sclerocytes.Aflatoxin B1 is one of the aflatoxins produced by the fungus Aspergillus flavus, which develops mainly in food. This fungus can develop in food, and its incidence is quite high in the storage of cereals, peanuts and cassava. Its development is rapid, reaching extremely high levels in favourable environments with ideal temperature and humidity (BIS, 2012). Caldas et al. (2002) states that it is not uncommon for people to be poisoned by mycotoxins in food and medicinal species in the form of tea. The use of plants for therapeutic purposes with no guarantee of safety and efficacy is aggravated by indiscriminate and uncontrolled trade, posing a danger to consumers due to the risk of acquiring and using poor quality products, as well as inappropriate post-harvest storage conditions (ARAÚJO; OHARA, 2000). The Rhizopus genus is also considered pathogenic because it is an opportunistic saprophyte (ESTEVES et al., 1990; BUGNO et al., 2005). Aspergillosis is a mycosis produced by opportunistic fungi of the genus Aspergillus, mainly A. fumigatus and less frequently by A. flavus, A. nidulans, A. niger, A. terrus and A. ochraceus (MEZZARI and CAUDURO, 1996). In addition, moulds caused by Fusarium sp., Aspergillus sp. and Penicillium sp. are also part of this range of contaminating agents, and are also producers of mycotoxins (ZARONI et al., 2004), posing a risk to the user of medicinal plants in the form of tea. Zygomycosis is an occasional disease with cutaneous, subcutaneous and systemic clinical manifestations caused by opportunistic fungi that become pathogenic, the Rhizopus oryzae, R. rhizoposiformis and R. microsporus species being pathogenic agents of this disease (MEZZARI and CAUDURO, 1996).

Microorganisms can contaminate vegetables during the pre- and post-harvest stages. In the pre-harvest stage, the main sources of contamination are soil, improperly composted fertilisers, contaminated irrigation water, water used to apply fungicides and insecticides, dust, insects, domestic and wild animals and human handling (ZARONI et al., 2004). Sources of post-harvest contamination include handling humans, inadequate or insufficient cleaning of harvesting equipment, dirty and contaminated transport packaging, animals, insects, dust, waste water, etc. contaminated washing, transport vehicles and contamination of equipment used in post-harvest processing, drying and storage, carried out improperly, without any type of control (BRACKETT, 1999; BEUCHAT, 2002; KALKASLIEF-SOUZA et al., 2009; SATOMI et al, 2005).

Matos (1998) emphasises the essential care that must be taken when handling plant materials, such as drying them properly, maintaining minimum hygiene conditions to avoid the development of mould and avoiding the material coming into contact with the ground, dust or even rain, which favours the development of microorganisms in medicinal plants. Silva et al. (1995) warns that consumers run the risk of fungal contamination of the final product at various commercial outlets (BUGNO et al., 2006). It is therefore necessary to establish appropriate hygiene and health control measures to guarantee the quality and safety of this type of product, from collection, storage and handling to the final product (home remedies or herbal medicines). In addition to these adjustments, the microbiological control of plant material and its derivatives must be stricter and more efficient on the part of the manufacturers and more demanded by the government bodies responsible for guaranteeing the safety, efficacy and quality of the products. It is therefore extremely necessary for consumers to follow legal prescriptions and for the products to contain information about their quality, avoiding the probable contamination of users of medicinal and aromatic plants (FARMACOPÉIA BRASILEIRA, 2005; OLIVEIRA et al, 1997; DINIZ et al, 1997; MATOS, 2000).

Characterisation and use of medicinal plants: chamomile, melissa, fennel, cinnamon, hibiscus and mint.

Chamomile (Chamomilla recutita L. Rauschert), from the Asteraceae family, is native to Europe and has been acclimatised in some regions of Asia and countries on the American continent, with the largest cultivation area. The state of Paraná stands out as the largest producer, with a cultivated area of 700 ha and production of around 260 tonnes, which meets 74% of national consumption, which is 350 tonnes (DALLA COSTA, 2001; CORREA JÚNIOR et al., 2003). Part of the chamomile consumed in the country is imported from Argentina, the world's main producer (CURIONI, 2004). Chamomilla recutita is a herbaceous, annual, aromatic plant with flowers gathered in chapters and is widely cultivated practically everywhere in the world (LORENZI, 2008). According to Lorenzi (2008), it is considered to be one of the oldest plants used in traditional European medicine and is now officially included in the Pharmacopoeias of almost all countries. It is used in both scientific and folk medicine, in the form of infusions and decoctions (boiled chapters), to facilitate the elimination of gas, combat colic and stimulate appetite. It also acts topically by applying compresses of the still warm infusion to the abdomen to treat colic in children.

Melissa oficinalis L, from the Verbanaceae family, in the Londrina region popularly known as lemon balm or melissa, is a sub-shrub of variable morphology, with an average height of 150 cm. Its branches are thin, whitish, arched, long and brittle, the leaves are entire, opposite, with serrated edges and an acute apex, 3-6 cm long, the flowers are a purplish-blue colour, gathered in capituliform axillary inflorescences with a short axis and variable size (LORENZI, 2008). According to Ming (1992), this species is cultivated and used throughout Brazil for its analgesic, antispasmodic, calming, sedative and cytostatic properties, and its therapeutic effects have already been scientifically proven. According to Scheffer (1992), little information is available on the agronomic aspect and there is a need for scientific studies to reveal the behaviour of the species medicinal plants when subjected to production techniques without affecting the therapeutic value of the plant, taking into account the fact that the active ingredients can change depending on cultivation techniques (MADUEÑO BOX, 1973). According to Lorenzi (2008), Pimpinella anisum L. belonging to the Apiaceae family, known as fennel, is native to Asia and cultivated in southern Brazil; it is considered aromatic, annual, erect, up to 50 cm high, has compound leaves of various shapes, split and white flowers, arranged in umbels. The fruits are achenes, which have a sweet flavour and a strong smell. They are also dried (mericarpos) and have been used since ancient times as a stimulant for digestive functions, to eliminate gas, combat colic, headaches and stimulate lactation. The use of tea is internationally approved as a simple remedy for colds, coughs and bronchitis, fever, colic, inflammation of the mouth and throat, poor digestion and loss of appetite (LORENZI, 2008).

The medicinal plant Cinnamomum zeylanicum Blume (cinnamon) belongs to the Lauraceae family, and its centre of origin is Sri Lanka and south-east India. It is an important spice and is used in the perfumery and beverage industries (JAYAPRAKASHA et al., 2003), cultivated in various regions of the world. It is also used in the pharmaceutical industry, as it provides health benefits due to its active ingredients (KHAN et al., 2003). According to Shan et al. (2005), chemical analyses show that this plant is rich in polyphenols and volatile oils. Lorenzi (2008) describes the cinnamon plant as an aromatic tree 6 to 12 metres high, with opposite, ovate and three-lobed leaves. The flowers are numerous and gathered in branched racemes arranged in terminal panicles, greenish-yellow in colour. Boorhem (1999) recommends its popular use for treating children's diarrhoea, flu, worms, toothache and bad breath. Pharmacological tests have shown that cinnamon essential oil has antibacterial activity against micro-organisms that cause illnesses in the respiratory system of the human body (LORENZI, 2008).

The hibiscus (Hibiscus sabdariffa L.) is an annual or biennial sub-shrub, erect, branched, with a stubby stem, 80-140 cm high and native to the African continent. It has alternate, long-petiolate leaves, entire at the base of the plant and 3 or 4 located at the apex, with toothed margins, 5-12 cm long; the flowers are solitary, axillary, with a yellow corolla and the fruits are capsules covered in hyssipids (LORENZI, 2008). According to Panizza (1997), hibiscus can be used therapeutically as a diuretic, laxative, stomachic, calming and antiscorbutic by using the calyxes in teas, soft drinks, jams and condiments. Mint (Mentha piperita) is used for medicinal and food purposes (LORENZI; MATOS, 2002). Its cultivation is of economic importance due to its ability to produce and store essential oil, the main constituent of which is menthol, used in oral hygiene products, pharmaceuticals, cosmetics and foodstuffs. Menthol also has high antifungal activity and antibacterial potential (SOUZA et al., 1991).

3. MATERIAL AND METHODS

The species used in the study were chosen on the basis of the plant organs that have secretory or storage structures for the active ingredients, such as leaves, bark, flowers and fruit. The plant species chosen for the phytosanitary quality analyses were: chamomile (Chamomilla recutita L. Rauschert), mint (Mentha piperita), fennel (Pimpinella anisum L), hibiscus (Hibiscus sabdariffa L.), cinnamon (Cinnamomum zeylanicum Blume) and melissa (Melissa officinalis L.).The plant materials were purchased from five commercial outlets in the municipality of Londrina: Feira livre (site 1); natural products shops (sites 2, 3 and 4) and a homeopathic pharmacy (site 5). The identified samples were then transported to the laboratory. The criteria for choosing these sites were: a populated area in the north of the municipality (open market, site 1); shops in the city centre of different social levels (natural products shops, site 2, site 3 and site 4) and a site also in the city centre specialising in natural remedies (homeopathic pharmacy, site 5). These were specific places where medicinal plants were sold. The Petri dishes used in the study were washed with 70% alcohol and sterilised at 170°C for 60 minutes (MENDONÇA et al., 2009). After sterilisation, the Petri dishes were taken to the flow chamber to place the BDA (potato dextrose agar) culture medium and insert 40 g of each plant species divided into four different spots on the dish. The plates were then sealed and left at a constant temperature of 25°C for seven days in the absence of light (BRASIL, 2003). Two replicates were carried out for each species and point of commercialisation (place where the material was collected).

After seven days, the Petri dishes were analysed. Slides were prepared and stained with a lactophenol blue solution for dyeing. and observed under an optical microscope to identify fungal microorganisms and to check for the presence or absence of bacterial formations. The Manual for the Sanitary Analysis of Seeds (BRASIL, 2009) was used to identify the fungal agents. With regard to bacterial formations, it was not possible to identify their genera and species, as the laboratory does not have the appropriate technical conditions, such as a scanning electron microscope, to carry out this identification, so they were only visualised when present in the samples.

4. RESULTS AND DISCUSSION

The identification of fungal species in the samples analysed is shown in Table 1. The presence of pathogens with the potential to be harmful to human health was verified, such as Aspergillus niger, Aspergillus ochraceus, Penicillium sp and Rhizopus sp. The fungi Alternaria sp and Nigrospora sp are not harmful to health, but only cause damage to the plant's physiology. The predominant pathogens in the samples of fennel (Pimpinella anisum) seeds were Rhizopus sp and bacterial formations (not identified). All the sites showed phytosanitary contamination for this medicinal species. Of the harmful fungi, only Aspergillus flavus was not identified in the samples collected. Mendonça et al. (2009) pointed out that fungi of the Rhizopus genus are not as harmful as those of the Aspergillus genus. With regard to bacterial contamination, 80% of the samples acquired were contaminated, which means that there is a high possibility of finding bacterial formations that are harmful to health.

Table 1: Presence of fungi and bacteria in fennel (Pimpinella anisum) seeds collected from five commercial outlets in the municipality of Londrina. 2012.

Pathogens Location%

	L1	L2	L3	L4	L5	pathogen** **
Alternaria sp*	X		X	X		60
Aspergillus flavus**						0
Aspergillus niger**	X				X	40
Aspergillus ochraceus**	X	X		40		
Curvalaria sp*			X			20
Fusarium sp**	X			X	X	60
Nigrospora sp**	X					20
Penicillium sp**	X	X			X	60
Rhizopus sp**	X	X		X	X	80
Bacteria***	X	X	X	X		80
% local*****	87,5	37,5	25,0	37,5	50,0	---

* Pathogen with harmful potential for human health.

** Pathogen with potential not harmful to human health.

*** No bacteria were identified.

**** Percentage of pathogen presence at the five marketing points.

***** Percentage of pathogens present at the point of sale.

L1: Free market. L2, L3, L4: Specific natural products shops. L5: Homeopathic pharmacy.

In the samples of melissa leaves (Melissa oficianalis L), shown in Table 2, the predominant mycoflora identified was made up of Fusarium sp and Aspergillus niger, fungi capable of producing mycotoxins The five sample collection points showed phytosanitary contamination, undoubtedly jeopardising the therapeutic use of this plant. The site with the highest percentage of contamination was site 1 (open-air market). As with fennel, 80% bacterial formation was detected in the melissa samples.

Table 2: Presence of fungi and bacteria on melissa leaves (Melissa officinalis L) collected from five commercial outlets in the municipality of Londrina. 2012.

Pathogens Location %

	L1	L2	L3	L4	L5	pathogen****
Alternaria sp*						0
Aspergillus flavus**	X					20
Aspergillus niger**	X		X	X		60
Aspergillus ochraceus**			X			20
Curvalaria sp*					X	20
Fusarium sp**	X	X	X	X		80
Nigrospora sp**	X	X				40
Penicillium sp**	X					20
Rhizopus sp**		X			X	40
Bacteria***	X	X		X	X	80
% local*****	75,0	50,0	37,5	37,5	25,0	---

* Pathogen with harmful potential for human health.

** Pathogen with potential not harmful to human health.

*** No bacteria were identified.

**** Percentage of pathogen presence at the five marketing points.

***** Percentage of pathogens present at the point of sale.

L1: Free market. L2, L3, L4: Specific natural products shops. L5: Homeopathic pharmacy.

Table 3 shows the phytosanitary contamination values obtained from the samples acquired at the five marketing points for cinnamon bark (Cinnamomum zeylanicum Blume). The only pathogen not identified was Curvalaria sp, a fungus that is not harmful to human health. The predominant pathogen in the samples was Aspergillus niger, although this fungus does not have the capacity to produce high percentages of aflatoxins, its presence in the samples jeopardises the use of this medicinal species in the form of tea. With regard to the presence of bacteria in the cinnamon samples, 60% of the samples showed bacterial formations.

Table 3: Presence of fungi and bacteria in cinnamon bark (Cinnamomum zeylanicum Blume) collected from five commercial outlets in the municipality of Londrina. 2012.

Pathogens Location %

	L1	L2	L3	L4	L5	pathogen****
Alternaria sp*	X					20
Aspergillus flavus**		X				20
Aspergillus niger**	X	X			X	60
Aspergillus ochraceus**		X				20
Curvalaria sp*						0
Fusarium sp**	X			X		40
Nigrospora sp**	X					20
Penicillium sp**	X			X		40
Rhizopus sp**			X			20
Bacteria***	X	X		X		60
% local*****	62,5	50,0	12,5	37,5	12,5	---

* Pathogen with harmful potential for human health.

** Pathogen with potential not harmful to human health.

*** No bacteria were identified.

**** Percentage of pathogen presence at the five marketing points.

***** Percentage of pathogens present at the point of sale.

L1: Free market. L2, L3, L4: Specific natural products shops. L5: Homeopathic pharmacy.

The results obtained from analysing the microbiological contamination of chamomile flowers sold at different points in Londrina are shown in Table 4. A high percentage of contamination was found to be caused by Aspergillus flavus, a pathogen capable of producing carcinogenic compounds called aflatoxins. Kneifel (2002) identified the presence of Aspergillus flavus in several samples of medicinal species in a microbiological safety experiment with medicinal species in São Paulo, SP. For the author, the high incidence of this pathogen is related to all the post-harvest events, such as processing, drying and storage of these materials redistributed to commercial outlets. This finding implies the phytosanitary disqualification of chamomile flowers marketed in Londrina for soothing purposes. All the pathogens targeted at health were identified in the chamomile samples purchased at the Londrina commercial outlets. Given this information, the use of this medicinal species as a tea is compromised in terms of phytosanitary quality due to the high aflatoxin-producing mycoflora found in the chamomile flower samples. As with cinnamon, the samples of chamomile flowers purchased at Londrina's retail outlets showed 60% of samples with visualised bacterial formations.

Table 4: Presence of fungi and bacteria in chamomile flowers (Chamomilla recutita L. Rauschert) collected from five commercial outlets in the city of Londrina. 2012.

Pathogens Location %

	L1	L2	L3	L4	L5	pathogen****
Alternaria sp*	X		X			40
Aspergillus flavus**	X	X	X	X	X	100
Aspergillus niger**	X	X		X		60
Aspergillus ochraceus**	X					20
Curvalaria sp*	X					20
Fusarium sp**	X		X			40
Nigrospora sp**	X			X		40
Penicillium sp**	X	X				40
Rhizopus sp**	X			X		40
Bacteria***	X	X	X			60
% local*****	87,5	62,5	37,5	50,0	12,5	---

* Pathogen with harmful potential for human health.

** Pathogen with potential not harmful to human health.

*** No bacteria were identified.

**** Percentage of pathogen presence at the five marketing points.

***** Percentage of pathogens present at the point of sale.

L1: Free market. L2, L3, L4: Specific natural products shops. L5: Homeopathic pharmacy.

The frequency of harmful pathogens identified in the hibiscus samples is high (Table 5). Of all the samples analysed, only the fungus Aspergillus ochraceus was not found or identified. This implies reduced reliability and phytosanitary safety for the use of this medicinal species by consumers, even for marketing point 3, where only fungi of the Rhizopus genus were found in the two repetitions. The predominant mycoflora in the samples acquired at the five Hibiscus sabdariffa L marketing points was 60 per cent, as in the fennel seed samples. As such, the consumption of hibiscus infusion is, in terms of microbiological safety, not recommended because, in addition to considerable bacterial formations, it contains several other genera of fungal pathogens that are dangerous to human health. Bacterial formations were found in 60% of the hibiscus samples.

Table 5: Presence of fungi and bacteria in hibiscus (Hibiscus sabdariffa L.) collected from five commercial outlets in the city of Londrina. 2012.

Pathogens Location %

	L1	L2	L3	L4	L5	pathogen** **
Alternaria sp*						0
Aspergillus flavus**					X	20
Aspergillus niger**	X			X		40
Aspergillus ochraceus**						0
Curvalaria sp*						0
Fusarium sp**	X					20
Nigrospora sp**		X			X	40
Penicillium sp**	X			X	X	60
Rhizopus sp**			X	X		40
Bacteria***	X	X		X		60
% local*****	50,0	25,0	12,5	50,0	37,5	---

* Pathogen with harmful potential for human health.

** Pathogen with potential not harmful to human health.

*** No bacteria were identified.

**** Percentage of pathogen presence at the five marketing points.

***** Percentage of pathogens present at the point of sale.

L1: Free market. L2, L3, L4: Specific natural products shops. L5: Homeopathic pharmacy.

The visualisation and identification of genera and species potentially harmful to human health in the mint leaf samples (Mentha piperita) are shown in Table 6. Like the medicinal species chamomile (Chamomilla recutita L. Rauschert), all the mint leaf samples purchased at the five different medicinal species marketing points showed microbiological contamination, even the two non-harmful fungi (Alternaria sp and Curvalaria sp). Table 6 also shows the high number of bacterial formations seen in the mint leaf samples, i.e. 80%, and only in the samples from marketing point 4 were no bacterial formations detected.

Table 6: Presence of fungi and bacteria on mint leaves (Mentha piperita) collected from five commercial outlets in the city of Londrina. 2012.

Pathogens Location %

	L1	L2	L3	L4	L5	pathogen****
Alternaria sp*			X		X	40
Aspergillus flavus**		X				20
Aspergillus niger**	X	X			X	60
Aspergillus ochraceus**	X	X		40		
Curvalaria sp*			X			20
Fusarium sp**					X	20
Nigrospora sp**					X	20
Penicillium sp**	X		X		X	60
Rhizopus sp**	X	X		X	X	80
Bacteria***	X	X	X		X	80
% local*****	62,5	50,0	37,5	12,5	75,0	---

* Pathogen with harmful potential for human health.

** Pathogen with potential not harmful to human health.

*** No bacteria were identified.

**** Percentage of pathogen presence at the five marketing points.

***** Percentage of pathogens present at the point of sale.

L1: Free market. L2, L3, L4: Specific natural products shops. L5: Homeopathic pharmacy.

The assessment of the phytosanitary quality of medicinal plants in Brazil, carried out by Fischer and Saito (1988), Fischer et al (1993), Santos et al (1995), Dallagnol and Alves (1998), Amaral (1999) and Araújo and Ohara (2000) showed that the main microorganisms involved in the contamination of such products are fungi, with a high occurrence of the genera Aspergillus and Penicillium. According to Bugno et al. (2006), the contamination of plant products by fungi is in some way related to inadequate harvesting and post-harvesting conditions. This situation is aggravated by consumers of plants for medicinal use who buy them in markets and open-air fairs, where the intense movement of people, together with inadequate packaging, exposes the plant material to heat, humidity, dust and other products such as foodstuffs. perishable, effectively contribute to the additional contamination of the material. The information discussed can be summarised in the graph below.

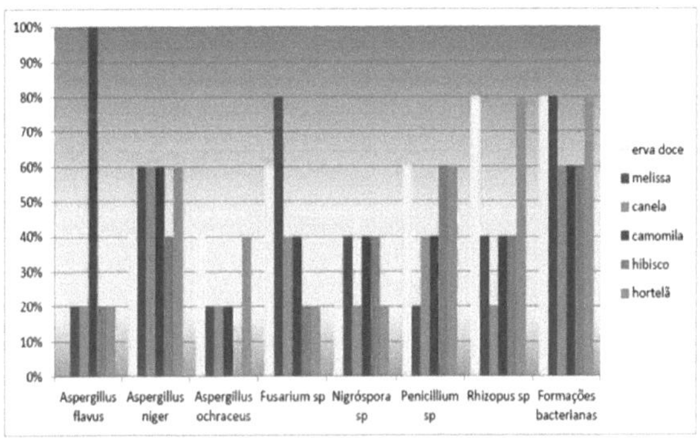

The graph shows the high level of phytosanitary contamination in the samples acquired at the points of sale of medicinal species in the municipality of Londrina, which represents a potential risk to tea users.

5. CONCLUSIONS

Chamomile (Chamomilla recutita L. Rauschert), mint (Mentha piperita), fennel (Pimpinella anisum L), hibiscus (Hibiscus sabdariffa L.), cinnamon (Cinnamomum zeylanicum Blume) and melissa (Melissa officinalis L.) bought in Londrina's shops are of poor phytosanitary quality.

REFERENCES

ABDOU, I. A.; ABOU-ZEID, A. A.; El-SHERBEENY, M. R.; ABOU-EL-GHEAT, Z. H., Antimicrobiol activities of Allium sativum, Allium cepa, Raphanus sativus, Capsicum frutescens, Eruca sativa, Allium kurrat on bacteria. **Quality of Plant Material,** p. 29-35, 1972.

ABIFISA-Brazilian Association of Companies Companies Sector Sector, **Supplements Food e Food Health** Available at: <http:/www.abifisa.org.br>. Accessed on 10 May 2012

AGRIOS, N.G. **Plant pathology**. 4. ed. San Diego: Academic Press, p.635., 1997.

ALBUQUERQUE,U.P. Quantitative ethnobotany or quantification in ethnobotany? **Ethnobotany Research and Applications**, 7: 1-2, 2009.

ALCORN, J.B. The scope and aims of ethnobotany in a developing world. In: R.E. Schultes and S.V. Reis (eds.). **Ethnobotany: evolution of a discipline**. Cambridge, Timber Press, p.23-39, 1995.

ALEXIADES, M.N.;SHELDON, J.W. **Ethnobotanical Research:** A Field Manual. Bronx, New York, The New York Botanical Garden, 1996.

ALMEIDA, S.P.; PROENÇA, C.E.B.; SANO, S.M.; RIBEIRO, J.F. **Cerrado:** Useful plant species. Planaltina, EMBRAPA-CPAC, Distrito Federal, p.464, 1998.ALVES S . A.M; FURTADO,G.Q;BERGAMIN FILHO, A . Influence of climatic conditions on soya bean rust. In: Zambolim L (Ed.). **Asian soya rust**. Viçosa MG. Suprema Gráfica e Editora Ltda. p. 37-59, 2006.

AMARAL, F. M. M. Frutos de Luffa operculata (L.) Cogn.: avaliação da comercialização e controle de qualidade de amostras adquiridas em mercados de São Luís/MA. **Brazilian Journal of Pharmacognosy.** São Luis. Federal University of Maranhão, p.27-31, 1999.

ARAÚJO, A. L. A.; OHARA, M. T. Qualidade microbiológica de drogas vegetais comercializadas em fairs de São Paulo e de infusos derivados. **Revista**

Brasileira de Ciências Farmacêuticas, v.36, n.1, jan./jun., p.129- 136, 2000.

BARBOSA, C.K.R.; COSTA, J.P.R.; BONFIM, F.P.G.; ALMEIDA, A.C.; MARTINS, E.R. Microbiological quality of medicinal plants grown and sold in Montes Claros, **Institute of Agrarian Sciences, Federal University of Minas Gerais Biotemas**, Minas Gerais, p. 77-81, 2010

BATALHA , M. O.; NANTES, J. F. D.; ALCÂNTARA, R. L. C.; MING, L. C.; CASTRO, D. M. de; LOURENZANI, A. E. B. S.; MACHADO, J. G. C. F.; RIBEIRO, P. M. T. Plantas medicinais no Estado de São Paulo: situação atual, perspectivas e obstraves ao desenvolvimento. **Revista Brasileira de Plantas Medicinais**, São Paulo (SP), p. 64-69, 2007.

BERG, M.E. **Plantas medicinais na Amazônia:** contribuição ao seu conhecimento sistemático 2 ed. Rev. E aum.- Belém: Museu Paraense Emílio Goeldi ,1993.

BEUCHAT, L. R. and GOLDEN, D. A., Antimicrobials occurring naturally in foods. **Food Technol.**,U.S.A, p. 134-142, 1989.

BEUCHAT, L. R. Ecological factors influencing survival and growth of human pathogens on raw fruits and vegetables. **Microbes and infection**, Paris, v. 4, p. 413-423, 2002.

BIS. **Internal Logistics Organisation Process Optimisation Utilities Energy Production Rationalisation Quality and Environmental Management Systems.** Available at: <http://www.bisbrazil.com.br/dicas/aflatoxina.htm. Accessed 05 Sep. 2012.

BOORHEM, R.L. **Secrets and Virtues of Medicinal Plants**, Reader Digest, Brazil, Rio de Janeiro, p.416, 1999.

BRACKETT, R.E. Incidence, contributing factors, and control of bacterial pathogens in produce. **Postharvest Biology and Technology**, Amsterdam, v. 15,p. 305-311, 1999.

BRANDÃO, M. G. L.; FREIRE, N.; VIANNA-SOARES, C. D. Surveillance of herbal medicines in Minas Gerais. Checking the quality of different commercial samples of camomile. **Cadernos de Saúde Pública**, v.14, n.2, p.693-700, 1998 a.

BRANDÃO, M. G. L.; FREIRE, N.; VIANNA-SOARES, C. D. Surveillance of herbal medicines in Minas Gerais. Verification of the quality of different commercial samples of camomile**. Cadernos de Saúde Pública**, v. 14, n. 3, p. 613-616, 1998 b.

BRAZIL. Ministry of Agriculture, Federal Government. **Technical Communication 101**, 1 edition, Brasilia, Federal District, 2003.

BRAZIL. Ministry of Agriculture, Federal Government. **Manual for the Sanitary Analysis of Seeds**, Brasília ,Distrito Federal, p.202, 2009

BRITO, V.F.S.; DANTAS, I.C.; DANTAS, G.D.S. Plantas medicinais utilizadas pela comissão de mulheres na zona rural no município de Lagoa Seca - PB, **Revista de Biologia e Farmácia**, vol3, n 1,Lagoa Seca, p.112-121, 2009.

BUGNO, A.; BUZZO, A.; NAKAMURA, C. T. Evaluation of microbial contamination in plant drugs**. Brazilian Journal of Pharmaceutical Sciences**. Vol. 41, n. 4, p.491-497, 2005.

BUGNO,A.;ALMODOVAR,A.A.B.;PEREIRA,T.C.;PINTO,T.J.A.; SABINO,M **Currence of toxigenic fungi in herbal drugs,** Instituto Adolfo Lutz, São Paulo, SP, Brazil; Faculdade de Ciências Farmacêuticas, Universidade de São Paulo, São Paulo, SP, Brazil,p.47-51, 2006

CALDAS, E. P.; SILVA, S. C.; OLIVEIRA, J. N. Aflatoxins and ochratoxins in food and risks to human health. **Revista de Saúde Pública**, São Paulo,v. 36, n. 3, 2002.

CALIXTO, J. B. Efficacy, safety, quality control, marketing and regulatory guidelines for herbal medicines (phytotherapeutic agents). **Brazilian Journal of Medical and Biological Research,** v. 33, p. 179-189, 2000.

CARRICONDE, C. **Introduction to the use of phytotherapics in PHC pathologies.**Olinda: Northeastern Centre for Popular Medicine, 2002.

CARVALHO, L.M.; COSTA, J.A.M.; CARNELOSSI, M.A.G. **Quality in**

Plantas Medicinais, Aracajú, p.56, Dec. 2010.

CASTELLANI, D. C, **Medicinal and aromatic plants:** non-timber forest products (NTFP) In. COELHO, M. F. B.; JÚNIOR, P. C.; DOMBROSKI, J. L. D. (eds.) **Diversos olhares em etnobiologia, etnoecologia e plantas medicinais**: anais do I Seminário Matogrossense e II Seminário Centro-Oeste de plantas medicinais- Cuiabá: Unicen, p.199-212, 2003.

CORREA JUNIOR, C.; MING, L. C.; SCHEFRER, M. C. **Plant cultivation medicinal and aromatic.** 2nd ed. Jaboticabal: FUNEP, p 162, 1994.

CORREA JÚNIOR, C.; EMPINOTTI, A.L.; SCHEFFER, M.C.; GRAÇA, L.R. **Study of the production chain of medicinal, spicy and aromatic plants.** Curitiba: EMATER-PR. p.26, 2003.

COUTINHO, D. F.; TRAVASSOS, L. M. A.; AMARAL, F. M. M. do. Study ethnobotany of medicinal plants used in indigenous communities in the state of Maranhão - Brazil. **Visão Acadêmica**, Curitiba, v. 3, n. 1, p. 7-12, Jan.-Jun./2002

CUMMINS, G.B. and HIRATSUKA, Y. **Illustrated Genera of Rust Fungi**. 3 ed. rev. Saint Paul:American Phytopathological Society, p. 240, 2003.

CURIONI, A.O. Cultivation and post harvest process for aromatic and medicinal plants in Argentina. In: **International symposium breeding research on medicinal and aromatic plants, 3, Latin American symposium on the production of medicinal and aromatic plants and condiments**, Campinas. Programme and Abstracts, Campinas: UNICAMP/IAC/CEFET/ UNESP, p.5, 2004.

DALLA COSTA, M.A. Process of agricultural production of chamomile in the municipality of Mandirituba, PR. (**Master's dissertation**), UFPR, Curitiba, p.69, 2001.

DALL'AGNOL, L. and ALVES, L. E. **Quality of aromatic plants commercialised in Curitiba-PR.** In: Symposium of Medicinal Plants of Brazil, 15, Águas de Lindóia, 1998. Programme and Abstracts. Águas de Lindóia,

National Council for Scientific and Technological Development, p.199, 1998.

DAVIDSON, P. M., Chemical preservatives and natural antimicrobial compounds, 29: 520-556. In: DOYLE, M. P.; BEUCHAT, L. R.; MONTVILLE, T. J. (eds), **Food Microbiology fundamentals and frontiers**, p.768, 1997.

DEFILIPPS, R. Conservation of Brazilian Medicinal Plants. In: **Biological Conservation Newsletter**. N.193, January 2001. Plant Conservation Unit. Department of Botany. Smithsonian National Museum of Natural History, 2001.

DEL PONTE, E.M.; GODOY, C.V.; Li. X; YANG, X.B. Predicting severity of Asian soybean rust epidemics with empirical rainfall models. **Phytopathology** 96:797-803, 2006.

DINIZ, M. F. F. M.; OLIVEIRA, R. A. G.; MALTA JUNIOR, A. **Das plantas medicinais aos fitoterápicos:** abordagem multidisciplinar. 2ª ed. João Pessoa: UFPB/CCS, p.131, 1997.

DUARTE, M.C.T. **Atividade antimicrobiana** de **plantas medicinais e aromáticas utilizadas no Brasil**, Universidade Estadual de Campinas Centro Pluridisciplinar de Pesquisas Químicas, Biológicas e Agrícolas DMB - Divisão de Microbiologia. Available at: <http:/www.multiciencia.unicamp.br/art05> Campinas, SP, 2006. Accessed on 29 April 2012

ELISABETSKY,E. New directions in ethnopharmacology**. Journal of Ethnobiology**, 6(1): 121-128, 1986.

ESPÍNOLA, E. B. **Safe and rational use of herbal products**. Revista Racine, p. 52-53, June 1997.

ESTEVES, J. A.; CABRITA, J. D.; NOBRE, G. N**. Mycologia médica**. 2nd Ed. Porto (Lisbon): Calouste Gulbenkian Foundation, p.341, 1990.

ETKIN, N.L. **Ethnopharmacology:** Biological and behavioural perspectives in the study of indigenous medicines**.** p. 149-158. In: T.M. Johnson and C.F. Sargent (eds.). **Medical anthropology:** A handbook of theory and method. New York, Greenwood Press, 1990.

FARMACOPÉIA BRASILEIRA. 4th ed. Brasília: Atheneu Editora São Paulo

Ltda, p.174, 2005.

FARNSWORTH, N. R.; AKERELE, O.; BINGEL, A. S.; SOEJARTO, S. S.; GUO, Z. G. **Medicinal plants in therapy**. Bulletin of the World Health Organisation, v. 63, n. 6, p. 965-981, 1985.

FERREIRA, S. H. Medicines from medicinal plants in Brazil. **Brazilian Academy of Sciences**, p. 133, 1998. FIGUEREDO, C.A. **Fitoterapia**, NEPHF, p. 156, 2006.

FISCHER, D. C. H.; SAITO, T. Microbial contamination in herbal medicines. **Journal of Pharmacy and Biochemistry.** USP, São Paulo, v. 24, n. 2, p. 143-144, 1988.

FISCHER, D. C. H.; OHARA, M. T.; SAITO, T. Microbial contamination in herbal medicines in solid form. **Journal of Pharmacy and Biochemistry**. USP, São Paulo, v.29,n.2, p.81-88, 1993.

FORD, R.I. Ethnobotany: historical diversity and synthesis. In: R.I. Ford; M. Hodge and W.L. Merril (eds.). **The nature and status of ethnobotany.** Annals of Arnold Arboretum**.** Michigan: Museum of Anthropology, University of Michigan, **Anthropological Papers** 67: 33-49, 1978.

GANDHI, D. N. and GHODEKAR, D. R., Antibacterial activity of garlic extract against lactic acid bacteria and contaminants of fermented milks. **Indian J. Dairy Science**. 41: 511-512, 1988.

HAMILTON, A.C.; SHENGJI, P.; KESSY, J.; KHAN, A.A.; LAGOS-WITTE, S.; SHINWARI, Z.K. **The purposes and teaching of Applied Ethnobotany**. Godalming, People and Plants working paper. 11, 2003.

HAWKSWORTH, D.L.; KIRK, P.M.; SUTTON, B.C.; PEGLER, D.N. **Ainsworth and Bisby's dictionary of the fungi**. 8. edition Wallingford:CAB international, p. 650, 1995.

HERRICK, J.W. The symbolic roots of three potent Iroquois medicinal plants. p. 134-155. In: L. Romanucci- Ross; D.E. Moerman and L.R. Tancredi (eds.). **The anthropology of medicine:** From culture to method. South Hadley, J.F.

Bergin ,1983.

HITOKOTO, H.; MOROZUMI; S., WAUKE, T.; SAKAI, S.; KURATA,H. Inhibitory effects of spices on growth and toxin production of toxigenic fungi. **Applied and Environmental Microbiology**, 39: 818-822, 1980.

JAYAPRAKASHA, G.K.; RAO L.J.; SAKARIAH, K.K. **Chemical composition of volatile oil from Cinnamomum zeylanicum**, p.57, 2003.

JORGE, S.S.A. **Plantas Medicinais**. Coletânea de Saberes, p.85, 2009.

KALKASLIEF-SOUZA, S. B.;KIKUCHI, I. S.; MANSANO, R. D. Microbial decontamination study of some medicinal plants by Plasma treatment. **Acta Horticulturae**, Wageningen, v. 826, p. 205-211, 2009.

KARL, A.C; SOUZA, R.M and MATTOS, J.K.A. Pathogenicity of Meloidogyne janica on four species of medicinal plants. **Horticultura Brasileira** p.118- 121, 1997.

KNEIFEL, W. Microbial contamination of medicinal plants. A review. **Planta Medica**, v68, p.5-15, 2002.

KHAN, A; SAFDAR, M.; KHAN, M.M.A.; KHATTAK, K.N.; ANDERSON, R.A. Cinnamom improves glucose and lipids of people with type 2 diabetes. **Diabetes Care**, p.26, 2003.

LIMA, C.S.;SOUZA, P.E. de; BOTELHO, A.O. Fungi of the family Pucciniaceae causing rust in medicinal plants. **Fitopatologia Brasileira** p.499-503. 2004.

LORENZI H.; MATOS, F.J.A. **Plantas Medicinais do Brasil: Nativas e Exóticas.** Instituto Plantarum, Nova Odessa, 2002.

LORENZI, H. **Plantas Medicinais no Brasil Nativas e Exóticas**, 2 Edição, p.544, 2008

MABROUK, S. S.; EL-SHAYEB, N. M. A.Inhibition of aflatoxin formation by some spices. Z. Lebensm. Unters. Forsch., 171: 344-347. In: GOLOB, P., MOSS, C., DALES, M., FIDGEN, A., EVANS, J.; GUDRUPS, I. (eds.). **The Use of Spices and Medicinals as Bioactive Protectants for Grains**. FAO

Agric. Serv. Bull., p.137, 1980.

MACIEL,M.A.M.; PINTO,A.C.; VEIGA JÚNIOR,V.F.; GRYBERG,N.F. **Medicinal plants:** the need for multidisciplinary studies. Federal University of Rio de Janeiro, Rio de Janeiro-RJ, Quim. Nova, Vol. 25, No. 3, p.429-438, 2002.

MADUEÑO BOX, M. **Cultivo de plantas medicinais**. Madrid: Labour, p.490, 1973

MARTIN, G.J. **Ethnobotany - A methods manual**. London, Ed. Chapman and Hall, p.268, 1995.

MARTINS, E.R; CASTRO, D.M. de; CASTELLANI, D.C.; DIAS, J.E. **Plantas Medicinals**. Viçosa: UFV, 1995.

MARTINS, E. R.; CASTRO, D.M.; CASTELLANI, D.C.; DIAS, J.E. **Plantas Medicinal**. Viçosa/ MG: UFV, p. 220, 2003.

MATOS, F. J. A; LORENZI, H. **Plantas Medicinais no Brasil - Nativas e Exóticas.** Nova Odessa, SP: Instituto Platarum, 2002.

MATOS, F.J.A. **Farmácias vivas**. 3rd Ed., Fortaleza: EUFC, 1998.

MATOS, F. J. A. **Plantas medicinais: guia de seleção e emprego de plantas usadas em fitoterapia no nordeste do Brasil**. 2ª ed. Fortaleza: Imprensa Universitária-UFC, p.344, 2000.

MEZZARI, A.; CAUDURO, P. F. **Mycology in the laboratory**. Editora Sagra, Brasil, Porto Alegre, p.139, 1996.

MELCHING, J.S; DOWLER, W.M; KOOGLE, D.L; ROYER, M.H. Effect of duration, frequency, and temperature of leaf wetness period on soybean rust. **Plant Disease** 73:117-122, 1989.

MELO, J. G.; MARTINS, J. D. G. R.; AMORIM, E. L. C. Quality of medicinal plant-based products marketed in Brazil: horse chestnut (Aesculus hippocastanum L.; lemon grass (Cymbopogon citratus (D.C.) Stapf.); and centella (Centela asiática (L.) Urban) **Acta Botânica Brasílica**, Brasília, v. 21, n. 1, 2007.

MENDONÇA, R.; FELFILI, J.M.; WALTER, B.M.T.; SILVA JÚNIOR, M.C.; REZENDE, A.V.; FILGUEIRAS, T.S.; NOGUEIRA, P.E.N. **Vascular Flora of the Cerrado**, p. 287-556, 1998.

MENDONÇA, M.B.; HIDALGO, A.F.; CHAVES, F.C.M. Isolating e identification of fungi with pathogenic potential for human health in medicinal plant material commercialised in Manaus. **Horticultura Brasileira**, p. 1208-1214, 2009.

MEYERS, N. Tropical moist forests; over-exploited and under-utilised? **Forest Ecology and Management** , 1983.

MING, L.C. Influence of different levels of organic fertiliser on biomass production and essential oil content of Lippia alba (Mill.) N.E.Br. -Verbenaceae. **Master's dissertation**. Curitiba: UFPR, p.206, 1992.

MING, L.C.; SILVA, S.M.P; SILVA, M.A.S.; HIDALGO, A.H; MARCHESE, J.A.; CHAVES, F.C.M. **Department of Plant Production Horticulture Sector.** Faculty of Agronomic Sciences UNESP - Botucatu - SP, 2009

MORGAN, R. **Encyclopaedia of herbs and medicinal plants.** São Paulo: Hemus, p.555, 1989.

NUNES, G.P.; SILVA, M.F; RESENDE, U.M.; SIQUEIRA, J.M. de.Plantas medicinal products sold by raizeiros in the centre of Campo Grande, Mato Grosso do Sul. **Brazilian Journal of Pharmacognosy**, p 82-83, 2003

OLIVEIRA, C. A. F.; GERMANO, P.M. Aflatoxinas: conceitos sobremecanismos de toxicidade e seu envolvimento na etiologia do câncer hepático celular. **Revista de Saúde Pública**, v. 31, n. 4, p. 417-424, 1997.

PANIZZA, S. **Plantas que curam.** São Paulo: IBRASA, p.120, 1997.

PLOTKIN, M. J. **Traditional knowledge of medicinal plants:** the search for new jungle medicines. In: AKERELE, O.; HEYWOOD, V.; SYNGE, H. **Conservation of medicinal plants**. Cambridge; Cambridge University Press, p. 53-64, 1991.

PRINCIPE, P.P. **The value of biological diversity among medicinal plants.**

Paris, Environment Directorate, Organisation for Economic Cooperation and Development, 1985.

PRUTHI, J. S., **Spices and condiments:** chemistry, microbiology, technology. Advances in food research, suppl. 4, Academic Press, New York, p.463, 1980.

RITTER, M. R.; SOBIERAJSKY, G. R.; SCHENKEL, E. P.; MENTZ, L. A. C. Plants used as medicinal plants in the municipality of Ipê. **Revista Brasileira de Farmacognosia**, RS, Brazil. vol.12, n. 2, p. 51-62, 2002.

ROITMAN, I.; TRAVASSOS, L. R.; AZEVEDO, J. L. **Treatise of microbiology**. São Paulo: Manole, p.186, 1988.

ROSA, F.O. Studies in diseases plant diseases. Available at: < http://fitopatologia1.blogspot.com.br/2010/04/aspectos-gerais-e-morfologicos-do-fungo_15.html>. Accessed on 5 September 2012.

RUSSOMANNO, O.M.R.; KRUPPA, P.C. Fungal diseases of medicinal, aromatic and spice plants. **Divulgação Técnica Biológica**, Instituto Biológico, Centro de Pesquisa e Desenvolvimento de Sanidade Vegetal, São Paulo (SP), p. 31-37, June, 2010.

SAAD, B; AZAIZEH, H; SAID, O. **Tradition and Perspectives of Arab Herbal Medicine:** A Review, 2(4): 475-479, october, 2005.

SANGUINETTI, E. E. **Plantas que curam.**2 ed. Editora Rígel, Porto Alegre p. 208, 1989.

SANTOS, P. R. V.; OLIVEIRA, A. C. X.; TOMASSINI, T. C. B. Control microbiology of herbal products**. Revista Farmacêutica Bioquímica.** University of São Paulo, v. 31,n.1, p.35-38, 1995.

SATOMI, L. C.; SORIANI, R. R.; PINTO, T. J. A. Decontamination of plant drugs using gamma radiation and ethylene oxide: microbial and chemical aspects. **Revista Brasileira de Ciência Farmacêuticas**, São Paulo, v. 41, n. 4, p. 445-450, 2005.

SCHEFFER, M.C. Script for the study of agronomic aspects of medicinal plants selected for phytotherapy by SUS-PR/ CEMEPAR. **SOB Informa** v. 10-11, n.

1-2, p. 29- 31, 1992.

SCHULTES, R.E. ;REIS, S.V. (eds.). **Ethnobotny: evolution of a discipline**. Cambridge, Timber Press, 1995.

SHAN, B; CAI, Y.Z.; SUN, M.; CORKE, H. Antioxidant Capacity of 26 spices extracts and characterisation of their phenolic constituents. **J Agriculture Food Chemist**, p.53, 2005.

SILVA, I.; FRANCO, S.L.; MOLINARI, S.L. **Notções sobre o organismo humano e utilização de plantas medicinais**. 3rd Ed., Cascavel (PR), 1995.

SOUZA, M.P.;MATOS, N.E.O.; MATOS, F.J.A. **Constituintes Químicos de Plantas Medicinais Brasileiras**. University Press/UFC, Fortaleza, 1991.

SOUZA, R.M.; MATTOS, J.K.A; KARL, A.C. Preliminary evaluation of the reaction o f medicinal plants to Meloidogyne javanica and M. incognita. **Horticultura Brasileira** vol.13: p. 209-211,1995.

SUZUKI, O. Market for herbal medicines in Brazil. In: SCHULZ, V.; HÄNSEL, R.; TYLER, V. E. **Revista Brasileira de Plantas Medicinais** 4. ed. São Paulo: Manole, p. 94- 99, 2002.

TORTORA, G. J.; FUNKE, B. R.; CASE, C. L. **Microbiology**. 6. ed. Porto Alegre: Artmed, 2000.

TSCHANZ, A.T. Soybean rust epidemiology: Final Report. **Asian Vegetable Research and Development Centre**, Shanhua, Taiwan, 1984.

URBEN, A.F.;MATTOS, J.K.A.; MENDES, M.A.S. Fungi associated with leaf spots on medicinal plants. **Brazilian Phytopathology** Distrito Federal, vol12: p. 390-394. 1987.

VEIGA JÚNIOR, F.V.; PINTO, A.C; MACIEL, M.A. **Plantas medicinais:** cura segura? Universidade Federal do Rio Grande do Norte, Campus Universitário, Natal - RN, vol. 28, no. 3, p. 519-528, 2005

ZARONI, M.; PONTAROLO, R.; ABRÃO, W. S. M. Microbiological quality of medicinal plants produced in the State of Paraná. **Revista Brasileira de Farmacognosia**, São Paulo, v. 14, n. 1, 2004.

APPENDICES

APPENDIX A

Illustrations of how the work was carried out.

Figure 1: Samples ofmedicinal acquired at commercial Londrina.

Figure 2: Clear difference between the types of packaging marketed for tea.

Figure 3: Petri dishes in the phytopathology laboratory.

Figure 4: Petri dishes with the results of phytosanitary contamination.

Figure 5: Alternaria sp and bacterial formations visualised on fennel (Pimpinella anisum)

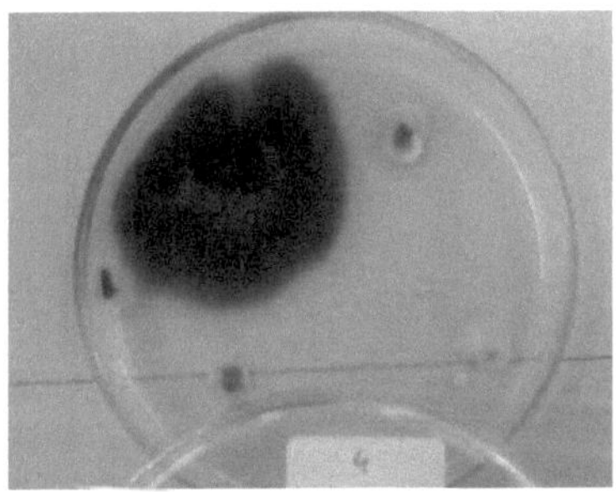

Figure 6: Aspergillus niger identified on melissa leaves (Melissa oficinalis L.).

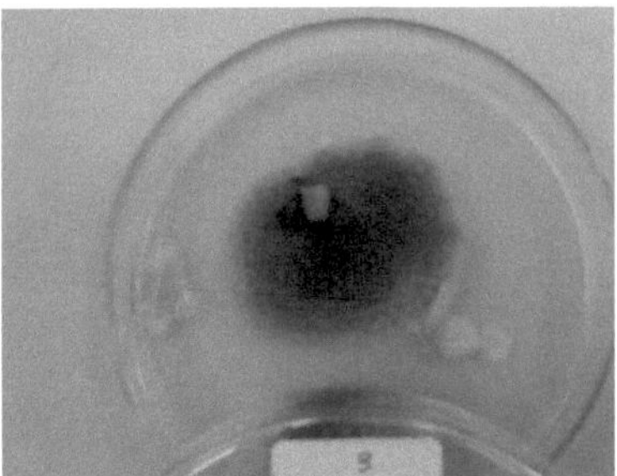

Figure 7: Bacteria visualised on cinnamon (Cinnamomum zeylanicum Blume).

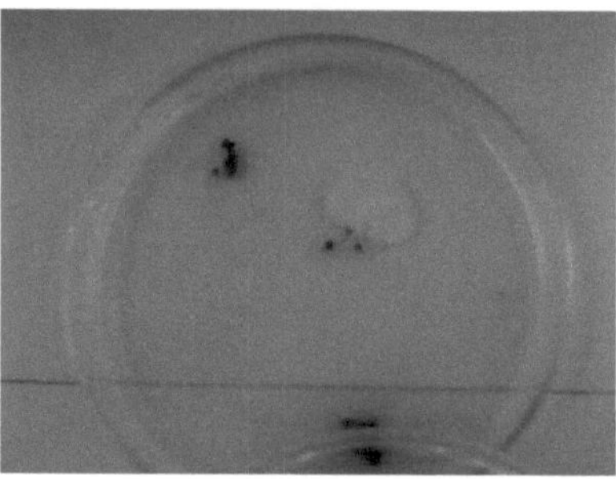

Figure 8: Aspergillus flavus observed in chamomile flowers (Chamomilla recutita L. Rauschert).

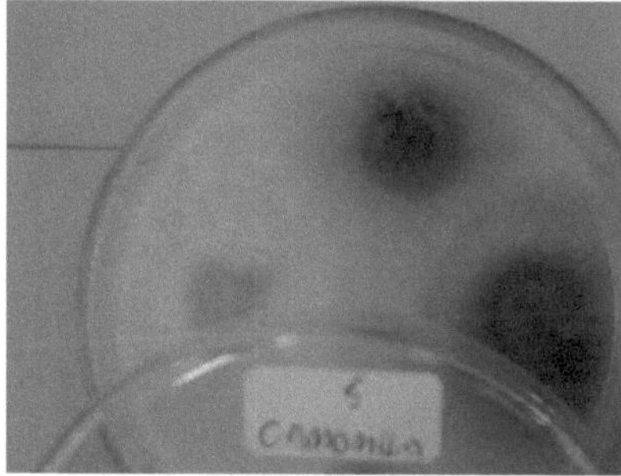

Figure 9: Penicillium sp observed on hibiscus (Hibiscus sabdariffa L.)

Figure 10: Aspergillus niger, Rhizopus sp and bacterial formations observed on mint leaves (Mentha piperita).

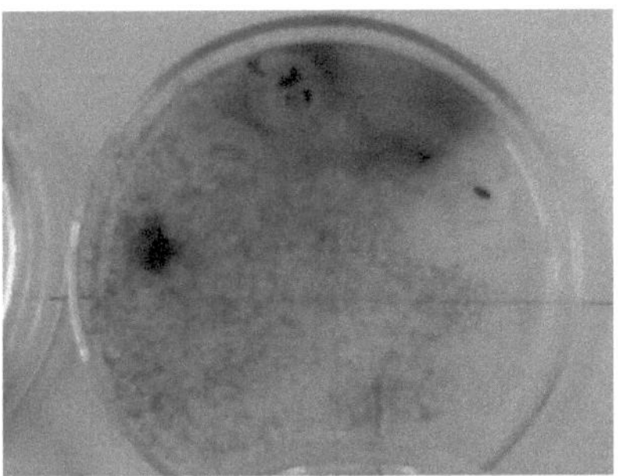

Printed by Books on Demand GmbH, Norderstedt / Germany